Name: Eng. / Mostafa Yacoub
Abdellatif Mahmoud

Nationality: Egyptian

ORCID: 0000-0002-9991-4624

Email:

moshhaabma2015@gmail.com

Qualification: civil engineer
Cairo University 2003

- **<u>Last digit of prime numbers:</u>**

  In this paper or research, we will prove that prime numbers (Except prime numbers 2 and 5) last digit have equal chances to be 1 3 7, or 9 based on my discovered formula that connects prime and composite numbers.

My discovered formula:

Definitions:

Array PTBP

It is the following Array of odd numbers

$$\begin{vmatrix} 1 & 3 & 7 & 9 \\ 11 & 13 & 17 & 19 \\ 21 & 23 & 27 & 29 \\ 31 & 33 & 37 & 39 \\ 41 & 43 & 47 & 49 \\ 51 & 53 & 57 & 59 \end{vmatrix}$$

And so on….

- **For a given set of consecutive primes whose numbers =n that start with prime 3 and end with prime F and not including prime 2 and prime 5**

i.e.

set=[3,7,11,13,…………………………………

………………………………..,F]

S=product of those consecutive primes

i.e

$$S = \prod_{i=3}^{i=F} (i)$$

Range=R_k = 10 × S×k

Where k = [1, 2, 3, 4, ...,
∞(infinity)

i.e R_1=10 x S x 1 and R_2=10 x S x 2

And so on

- Number of composite numbers that belong to Array PTBP and created by the effect of those consecutive primes within the range R_K

- =[(K × 4$^{× \frac{S}{3}}$) + (

$$\sum_{j=7}^{j=F} (K \times 4 \times \left(\frac{S}{j}\right) \times$$

$$i = prime\ number\ befor\ current\ prime\ number\ j$$

$$\prod_{i=7} \qquad \left(\frac{i-1}{i}\right)$$

$$)]-(n)$$

Where j =consecutive values of primes

7, 11, 13,.............., F

And i= consecutive values of primes

3, 7, 11, 13,........, prime before current j prime

The previous formula can be applied for any number of consecutive prime numbers that start with prime number 3

- The first term $^{(k \times 4 \times \frac{S}{3})}$ represents the count of unique Composite numbers +1 that belong to the Array PTBP and are created by prime number 3 within the range

$$R_k = 10 \times S \times k$$

- The second term

$$\sum_{j=7}^{j=F} (K \times 4 \times (\frac{S}{j}) \times$$

$i = $ *prime number befor current prime number j*

$$\prod_{i=7} \qquad (\frac{i-1}{i})$$

Represent the count of unique Composite numbers+n-1 that belong to the Array PTBP and are created by each prime number after the prime number 3 within the range

R_k =10 $\times$ S $\times$ k

- **The third term (-n)**

Subtracting n (number of consecutive primes starting from prime number 3) because the count of composite numbers generated from those consecutive primes includes the count of those primes in the range

$R_k = 10 \times S \times k$

- Explanation and proof for my theory in my previous paper (prime number theory)
- We will mention only the concept of number cycle

  We can use the number cycle concept to understand the behavior of consecutive primes in creating composite numbers.

  i.e.

  $$S = \prod_{i=3}^{i=F} (i)$$

  Range=cycle range= $R_k = 10 \times S \times k$

  Where k= [1, 2, 3, 4, ..., ∞(infinity)

  i.e. R_1=10 x S x 1 and R_2=10 x S x 2

  And so on

- Now consider only one k value =1

- **For a set of consecutive primes and according to my formula the result will be**

$$= [(K \times 4^{\times \frac{S}{3}}) + ($$

$$\sum_{j=7}^{j=F} (K \times 4 \times (\frac{S}{j}) \times$$

$$i = prime\ number\ before\ current\ prime\ number\ j$$

$$\prod_{i=7} \qquad (\frac{i-1}{i})$$

$$)] - (n)$$

And

including the count of prime numbers within the set

$$=[(K \times 4^{\times \frac{S}{3}}) + ($$

$$\sum_{j=7}^{j=F} (K \times 4 \times (\frac{S}{j}) \times$$

$i = prime\ number\ befor\ current\ prime\ number\ j$

$$\prod_{i=7} \qquad (\frac{i-1}{i})$$

$$)]$$

Which represent the count of numbers (that belong to the Array PTBP) that is divisible of the prime numbers that belong to the set of consecutive primes

- **Considering the term of prime number three gives a count**

$$= \frac{(4 \times \frac{S}{3})}{}$$

- **The values of that count are equivalent to a part of the array**

PTBP with the first (S/3) rows multiplied by prime number three

- **For example if S= 3 $\times 7$**

Then we have the following array that represents the values of that count 4 $^{\times \frac{S}{3}}$

With number of rows =21 / 3 = 7 rows

$$1, \quad 3, \quad 7, \quad 9$$
$$11, 13, 17, \quad 19$$
$$21, 23, 27, \quad 29$$
$$31, 33, 37, \quad 39$$
$$41, 43, 47, \quad 49$$
$$51, 53, 57, \quad 59$$
$$61, 63, 67, \quad 69$$

- **Then multiply each element by 3 then we have**

3,	9,	21,	27
33,	39,	51,	57
63,	69,	81,	87
93,	99,	111,	117
123,	129,	141,	147
153,	159,	171,	177
183,	189,	201,	207

- **So we must have a uniform distribution of the last digit**

- **And now we will study the second part**

$$\sum_{j=7}^{j=F=lastprime} (4 \times \left(\frac{S}{j}\right) \times \prod_{i=3}^{i=primebeforecurrentjprime} (\frac{i-1}{i}))$$

- **The same proof is valid for prime7,11, and so on**
  But we have a problem with the intersection of throw out result of prime number j with a previous prime number within the specified set of consecutive prime numbers
- **And now we will prove that all those values of intersections have a uniform distribution for**

the last digits 1 and 3 and 7 and 9

- set=[3,7,11,13,..........................,F]

S=product of those consecutive primes

$$S =\prod_{i=3}^{i=F} (i)$$

We have the following sequence

- $S/3 =\prod_{i=7}^{i=F} (i)$

- $S/7= 3 \times \prod_{i=11}^{i=F} (i)$

- $S/(3 \times 7)$

- $S/(3 \times 11)$
- $S/(7 \times 11)$
- $S/(3 \times 7 \times 11)$

- $S/(13)$
- $S/(3 \times 13)$
- $S/(7 \times 13)$
- $S/(3 \times 7 \times 13)$
- $S/(13 \times 11)$
- $S/(3 \times 13 \times 11)$
- $S/(7 \times 13 \times 11)$
- $S/(3 \times 7 \times 13 \times 11)$
- $S/(17)$
- $S/(3 \times 17)$
- $S/(7 \times 17)$
- $S/(3 \times 7 \times 17)$

- S/(17 × 11)
- S/(17 × 11 × 3)
- S/(17 × 11 × 7)
- S/(17 × 11 × 3 × 7)
- S/(17 × 13)
- S/(17 × 13 × 3)
- S/(17 × 13 × 7)
- S/(17 × 13 × 3 × 7)
- S/(17 × 13 × 11)
- S/(17 × 13 × 11 × 3)
- S/(17 × 13 × 11 × 7)
- S/(17 × 13 × 11 × 3 × 7)
- S/(19)
- S/(3 × 19)
- S/(7 × 19)
- S/(3 × 7 × 19)
- S/(19 × 11)
- S/(19 × 11 × 3)

- S/$(19 \times 11 \times 7)$
- S/$(19 \times 11 \times 3 \times 7)$
- S/(19×13)
- S/$(19 \times 13 \times 3)$
- S/$(19 \times 13 \times 7)$
- S/$(19 \times 13 \times 3 \times 7)$
- S/$(19 \times 13 \times 11)$
- S/$(19 \times 13 \times 11 \times 3)$
- S/$(19 \times 13 \times 11 \times 7)$
- S/$(19 \times 13 \times 11 \times 3 \times 7)$
- S/(19×17)
- S/$(19 \times 17 \times 3)$
- S/$(19 \times 17 \times 7)$
- S/$(19 \times 17 \times 3 \times 7)$
- S/$(19 \times 17 \times 11)$
- S/$(19 \times 17 \times 11 \times 3)$
- S/$(19 \times 17 \times 11 \times 7)$
- S/$(19 \times 11 \times 3 \times 7)$

- $S/(19 \times 17 \times 13)$
- $S/(19 \times 17 \times 13 \times 3)$
- $S/(19 \times 17 \times 13 \times 7)$
- $S/(19 \times 17 \times 13 \times 3 \times 7)$
- $S/(19 \times 17 \times 13 \times 11)$
- $S/(19 \times 17 \times 13 \times 11 \times 3)$
- $S/(19 \times 17 \times 13 \times 11 \times 7)$
- $S/(19 \times 17 \times 13 \times 11 \times 3 \times 7)$
- **We can continue in the same sequence until finishing all primes within that set**
- **So all those counts have corresponding values that represent several raw of array PTBP**
- **So all possible intersections have a uniform distribution of the last digit 1 3 7 9**

- **And so on for all prime numbers that belong to that set**
- **So the result from my formula has a uniform distribution of the last digit 1 3 7 9**

$$(4 \times S) - (4 \times \prod_{i=3}^{i=F} (i-1)) =$$

$$[[(4 \times \frac{S}{3}) + [$$

$$\sum_{j=7}^{j=F=lastprime} (4 \times (S/j) \times \prod_{i=3}^{i=prime\ before\ current\ j\ prime} (\frac{i-1}{i})$$

$$)]]]$$

*the count of t*he   **complementally**

part

$$= (4 \times \prod_{i=3}^{i=F} (i-1))$$

Which represent the count of numbers (that belong to the Array PTBP) that is not divisible by each prime number that belong to the set of consecutive primes has a uniform distribution of the last digit 1 3 7 9

- So prime numbers (Except prime numbers 2 and 5) last digit has equal chances of being 1 3 7 or 9 based on my discovered formula that connects prime and composite numbers.
- And now we will prove it but by mathematical equations
- For a set of consecutive primes Set=[3]
  We have the following array
  1   3   7   9

11 13 17 19
21 23 27 29

- The sum of each column =
  33 39 51 57 respectively
- We will name the result as S_{ij} where i represent the last prime of the set and j represents the column last digit 1 or 3 or 7 or 9
  S_{31} S_{33} S_{37} S_{39} respectively
- And now if we want the sum of columns for the cycle that represents a set of consecutive primes =[3,7]
  So let $S_7 = S_{71}+S_{73}+S_{77}+S_{79}$

$$= 20 \times \prod_{i=3}^{i=F} (i\char94 2)$$

- **More detail exist in my paper (prime number theory part 2)**
- **P=7 (p is the next prime number after prime number F)**

$$S_{71}=(S_{31} \times P) + (5 \times (\prod_{i=3}^{i=F} (i\char`^2)$$

$$) \times P \times (P-1))$$

$$= (S_{31} \times P) + ((20/4) \times (P\char`^2) \times ( \prod_{i=3}^{i=F} (i\char`^2)$$

$$)) - ((20/4) \times (\prod_{i=3}^{i=F} (i\char`^2)$$

$$) \times P\char`^2 / P)$$

$$=(S_{31} \times P) + ((1/4) \times S_7)-((1/4) \times S_7/P)$$

$$S_{71}=(S_{31} \times P) + ((S_7/4) \times ((P-1)/P))$$

$$S_{71}/\left(\prod_{i=3}^{i=P} (i) \right) = \left(S_{31} \, / \right.$$

$$\left. \prod_{i=3}^{i=F} (i) \right) + \left(5 \times \prod_{i=3}^{i=P} (i) \right) \times$$

((P-1)/P))

Last digit of the part $S_{71}/\left(\prod_{i=3}^{i=P} (i) \right)$ must be 1

And also the last digit of the

part $\left(S_{31} \, / \prod_{i=3}^{i=F} (i) \right)$ must be 1

(because $\prod_{i=3}^{i=F} (i)$ x the average value of the numbers of the first column =

summation of the numbers of that column within the Array PTBP produced from that particular set of consecutive primes)

- so the last digit of part 5 $\times$

$$\prod_{i=3}^{i=P} (i)$$

must be 5

- so the last digit of 5 $\times$ (p-1) must = 0
- i.e the last digit of prime P has equal chances of being 1 or 3 or 7 or 9